GABRIEL JEAN REMY DIATTA

Meteorology in the service of health

GABRIEL JEAN REMY DIATTA

Meteorology in the service of health

Epidemiological modeling of malaria (Plasmodium falciparum parasite) in the Ziguinchor region

ScienciaScripts

This book is a translation from the original published under ISBN 978-620-6-72483-4.

Publisher:
Sciencia Scripts
is a trademark of
Dodo Books Indian Ocean Ltd. and OmniScriptum S.R.L publishing group

120 High Road, East Finchley, London, N2 9ED, United Kingdom
Str. Armeneasca 28/1, office 1, Chisinau MD-2012, Republic of Moldova, Europe
Printed at: see last page
ISBN: 978-620-8-20874-5

EPIDEMIOLOGICAL MODELLING OF MALARIA (PARASITE: PLASMODIUM FALCIPARUM) IN THE ZIGUINCHOR REGION MR GABRIEL JEAN R. L. DIATTA, FEBRUARY 2024.

TABLE OF CONTENTS

SUMMARY

Malaria is a major health problem in Senegal. Given the seasonal nature of the disease, the introduction of decision-support tools such as the use of models to predict the occurrence of malaria epidemics would be of vital importance in the fight against the disease. This study therefore contributes to a better understanding of the relationship between climate and malaria, with a view to improving malaria surveillance and management. The study was carried out in the Ziguinchor region using data covering an eight (08) year period, from 2015 to 2022. Two types of monthly data were used, namely meteorological data and health data (number of confirmed cases of malaria). The methodology adopted consisted of graphical analysis, simple linear regression and multiple linear regression. The most relevant parameters were selected using the backward elimination method. Finally, the model was validated using four (04) statistical tests (the Shapiro-Wilk test, the variance inflation factor test, the Durbin-Watson test and the Goldef-Quandt test), followed by a robustness check using the Taylor diagram. The period favourable to the occurrence of malaria is characterised by the onset of winter, an average relative humidity ranging from 67 to 90%, a minimum temperature of over 22.4°C, a relatively low maximum temperature of under 33.3°C, a wind speed of between 0.78 and 1.5 m/s and a sunshine duration of between 9.1 and 10.8 hours. Malaria was significantly correlated with monthly maximum temperature (r= -0.66, p*), monthly mean relative humidity (r=0.67, p*) and monthly wind speed (r=-0.86, p**). Nevertheless, precipitation, maximum temperature and minimum temperature are the most relevant meteorological parameters in the model. The model is valid, reproducing the seasonality of the disease fairly well, but has a poor predictive capacity, tending to underestimate the number of malaria cases during the rainy season by up to 40%.

Key words: modelling, malaria, Senegal climate.

INTRODUCTION

Numerous studies have been undertaken in an attempt to clarify the relationship between climate and human, plant and animal pathologies. Variations in temperature, rainfall and ambient air humidity could therefore influence the evolution of ecosystems (Passike et al., 2021). This would significantly affect the development cycle of pathogens or biological vectors and encourage their proliferation. In fact, the idea that human health and disease are linked to climate goes back to antiquity. The Greek physician Hippocrates recounted that "epidemics are linked to the seasonal variability of the climate, so whoever wants to apply himself properly to medicine should take into account the different seasons of the year and their effects on human health, bearing in mind his geographical position" (Jouanna, 2020). Sometimes the climate intervenes through the opportunities it offers for the pathogen, its vector or a possible intermediate host to develop and proliferate. Sometimes it acts directly on the human body, weakening its natural defences. At other times, its role is mediated by the availability of foodstuffs and by nutritional status (Besancenot et al, 2004). Epidemiological studies have long suggested that meteorological factors, generally temperature, humidity and wind, can influence the incidence of infectious diseases (Xiao et al., 2020). During the rainy season, from June to October, diseases linked directly or indirectly to water such as cholera, malaria, filariasis and Nile Valley fever occur (Penguin, 1978). During the dry season, which is mainly marked by epidemics of meningitis, the climate is characterised by frequent dust-laden easterly winds (harmattan). This is compounded by a significant drop in humidity levels and a wide daily temperature range.

In West Africa, and more specifically in Senegal, it has been observed that over the years, peaks in malaria epidemics regularly occur in the wet season. In terms of epidemiological monitoring, a renewed focus on climate in malaria prevention would serve to recognise both the profoundly social nature of this disease and the necessarily holistic nature of our response. Seasonal trends in malaria notifications have been observed in several countries. Explanatory factors may include weather conditions, indoor overcrowding, environmental factors, etc.According to WHO (2020), malaria is the world's deadliest infectious disease, and is a major health problem in Senegal, where it is endemic with a seasonal upsurge, accounting for around 35% of consultations. Malaria has a significant impact on the health of the population, causing negative repercussions on the country's economy due to the reduced productivity of a population affected by the disease, to which must be added the risks of pollution when insecticides are sprayed, which can be toxic for ecosystems and lead to the appearance of resistant insects (Zongo, 2009). It has been observed that the disease is present in all regions of Senegal, with disparities in distribution and throughout the year, with periods of high and low contamination. Specifically, Ziguinchor, Senegal's natural lungs, is located in the climatic red zone of maximum suitability for malaria transmission (WHO, 2019). Given these considerations, the introduction of decision-making tools such as models capable of combining climatological, environmental and epidemiological data and information to predict the occurrence of malaria epidemics would be of vital importance in the fight against the disease.This study will therefore contribute to a better understanding of the relationship between climate and malaria, with a view to improving the surveillance and management of the disease in the Ziguinchor region. Specifically, the study aims to assess the statistical links between each meteorological parameter and malaria; to determine the meteorological characteristics of the period favourable to the occurrence of malaria; to identify the relevant meteorological parameters; and to develop an epidemiological model for forecasting malaria based on the relevant meteorological parameters.

CHAPTER I
MATERIAL AND METHODS

1.1 Justification for the choice of study area

The study area is a border region between Senegal-Guinea Bissau and Senegal-Gambia. The Ziguinchor region is located in the climatic red zone of maximum suitability for malaria transmission and has a very rich variety of epidemiological facies, making it a very interesting site for our study. The region was chosen because of its intense market gardening activity, which results in numerous watering points, and because it includes neighbourhoods with a high potential for An. gambiae breeding sites (market gardens, rice fields, puddles). It should also be remembered that the Ziguinchor region has been pre-eliminating malaria for several years (WHO, 2020).

1.1.1 Location

The Ziguinchor region lies between latitudes 12°33'N and 13°16'N and longitudes 15°90'W and 16°80'W (Figure 4), with a magnetic declination of 13°05. It covers an area of 7,339km2, or 3.73% of the national territory. It is bordered to the north by the Republic of the Gambia, to the south by the Republic of Guinea Bissau, to the east by the regions of Kolda and Sédhiou and to the west by the Atlantic Ocean (ANSD ,2019).

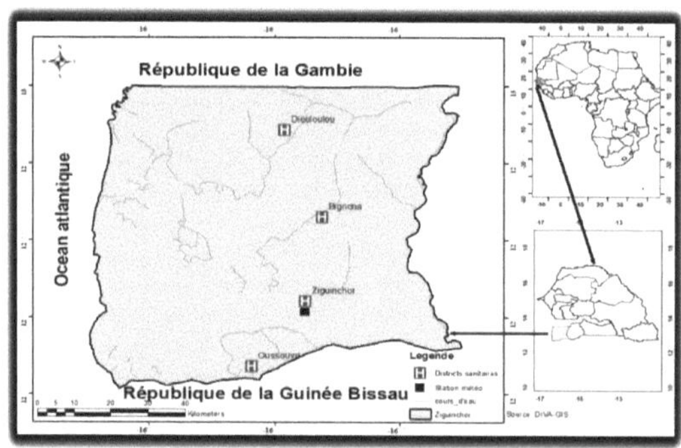

Figure 1: Map of the Ziguinchor region

1.1.2 Geography

1.1.2.1 Landforms and soil types

The region's terrain is generally flat. Along the Casamance river, the level is more or less that of the sea. A small stretch of coastline consists of of low-lying land and is less than a metre above sea level, which facilitates marine intrusion in this area.

The main types of soil found within the regional perimeter are :

• The hydro morphic soils in the valleys are used for rice growing and market gardening;

• Tropical ferruginous and ferralitic sandy or sandy-clay soils on plateaux and terraces forming watersheds, used for rainfed crops (groundnuts, cowpeas, rice, etc.) and colonised by woody formations, most often palm groves (ANSD,2019).

1.1.2.2 Vegetation and fauna

The region is subject to the influence of the sub-Guinean climate, which favours high rainfall compared with the central and northern regions of the country. We note the formation of a forest domain made up of dense dry forests and gallery forests located mainly in the southern part. Mangrove swamps and palm groves colonise the fluviomaritime zone, and there are also roast plantations. The region has significant wildlife potential. The forest galleries and certain classified forests contain a variety of animal species such as harnessed guibs, red-sided duikers, yellow-backed duikers and cercopithecines (green monkeys, patas and colobus monkeys), porcupines and reptiles. The rocky vegetation, which is so well represented, is the habitat of choice for green monkeys. In the department of Oussouye, and more specifically in Santhiaba-Manjaque, the Lower Casamance National Park is an important wildlife retreat area (ANSD,2019).

1.1.2.3 Hydrographic

The region's hydrographic network consists mainly of the Casamance River (a semi-permanent river that flows from June to March). This river is fed by the Soungrougrou, a 140km tributary, and the marigots of Guidel, Kamobeul, Bignona, etc. The surface area of the basin drained is around 20,150 km², including the major sub-basins (Baïla : 1,645 km², Bignona: 750 km², Kamobeul: 700 km², Guidel: 130 km² and Agnack: 133 km²) with highly variable volumes ranging from 60 to 280 million m3 /year. The Casamance River, 350 km long, is often bordered by mangroves and invaded by marine waters up to 200 km from its mouth (Diana Malari/Sédhiou), where it discharges highly variable volumes: 60 to 280 million m3 of water per year (ANSD, 2019).

1.1.3 Climatology

The Ziguinchor region has one of the highest rainfall rates in the country (ANSD, 2019). Normal rainfall is 1282.2 mm and there are two seasons: a wet season from June to October and a dry season from November to May, according to the 1991-2020 normal. Rainfall peaks in August (Figure 5). The entire region has a hot, tropical savannah climate that is more or less dry (ANSD, 2019). The average normal temperature is 26.5°C, peaking at 30°C in May. Low temperatures are recorded in December, January and February. Temperatures are mild during the rainy season (Figure 6).

In general, there is a relative uniformity in the duration of sunshine, which is significant throughout the year, with a slight drop in winter due to cloud cover (ANSD ,2019).

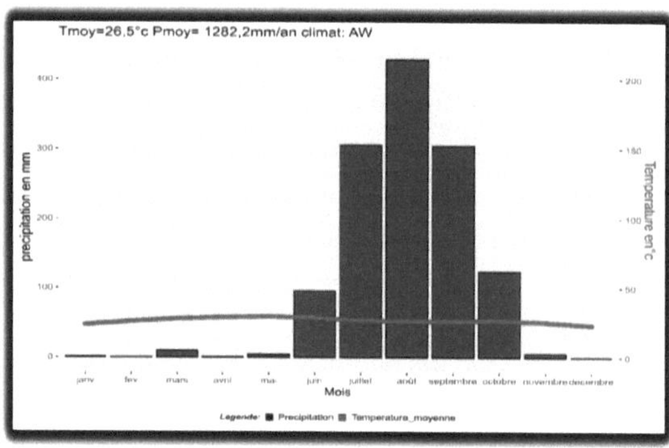

Figure 2: Umbrothermal diagram for the Ziguinchor region

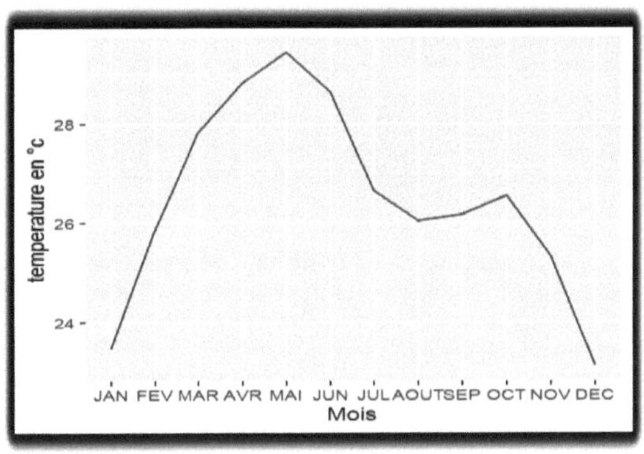

Figure 3: Monthly variation in mean temperature

Rainfall in the region is highly variable, with alternating years of deficit and surplus. Overall, rainfall in the Ziguinchor region varies from deficit to surplus.

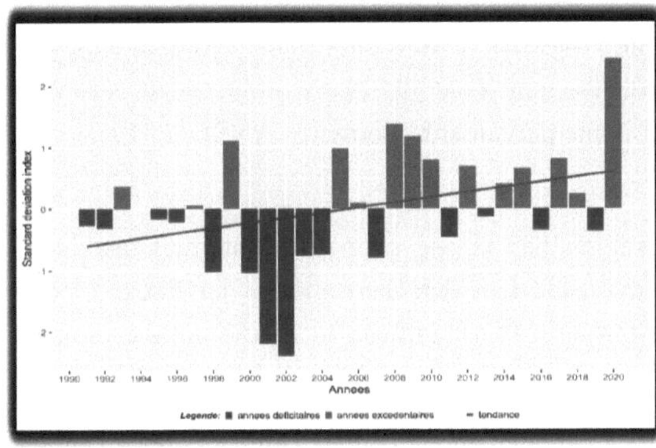

Figure 4: Rainfall index for the 1991-2020 normal

In terms of average temperature, the region has a high degree of thermal variability, with alternating hot and cold years. The overall trend in normal temperature for the Ziguinchor region varies little, although it does go from colder to warmer than normal.

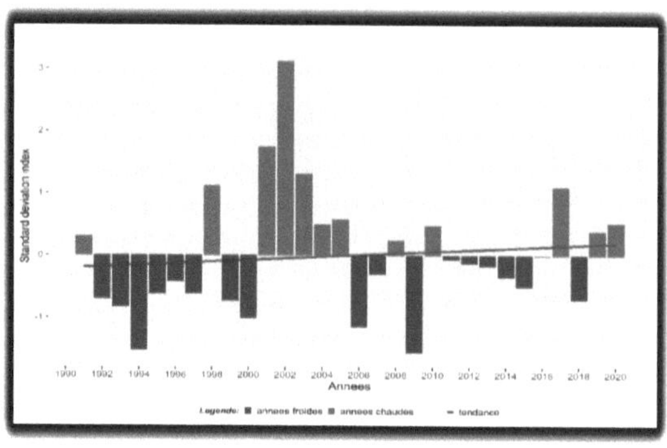

Figure 5: Thermal index for the 1991-2020 normal

1.1.4 Population

The population of the Ziguinchor region has more than doubled in the 40 years between 1976 and 2019, rising from 242,980 to 662,179. The majority of this population is young, over 70%. Men outnumber women, with a sex ratio of 106 men for every 100 women. Overall, The population density of the Ziguinchor region is 90 inhabitants per square kilometre (ANSD, 2019).

1.1.5 Economy

The Ziguinchor region has strong economic potential that is conducive to its emergence. However, the fact that the region is landlocked, combined with the crisis it is going through, is a handicap to harmonious economic development. The main economic activities are agriculture, fishing and tourism. The latter has experienced problems in recent years as a result of the crisis in the region (ANSD, 2019).

1.1.6 Health

In 2019, there were 237 public and semi-public health facilities in the Ziguinchor region. The number of facilities remained the same as in 2018. The majority of these facilities are health posts, which account for 51%, followed by health centres, which account for 41% of all public and semi-public health facilities in the region.There are only two primary care hospitals in the region, in the department of Ziguinchor and more specifically in the commune of Ziguinchor. There are 5 health centres, 3 of which are in the department of Bignona; the departments of Ziguinchor and Oussouye each have one health centre. The number of public and semi-public health facilities did not change between 2018 and 2019. The region also has a number of non-hospital public health establishments. These include the Pharmacie Régionale d'Approvisionnement (PRA), which specialises in essential medicines and products. This regional pharmacy plays a very important role in supplying medicines to other public and private pharmacies in the Ziguinchor region (ANSD, 2019).

1.2 Hardware

The material used consists of a meteorological file of monthly climatic parameters for each year (2015 to 2022) from the Ziguinchor synoptic weather station. The climatic parameters in question are : Wind speed (WIND), average relative humidity (Umoy), minimum (Tn) and maximum (Tx) temperature, precipitation (rain) and sunshine duration (INSO). The data used for the climatic parameters are the cumulative monthly rainfall data and the monthly averages for the rest of the parameters. A medical file (monthly number of malaria cases from 2015-2022). This health data on the number of cases comes from the monitoring and evaluation department of the national malaria control programme (PNLP) in the medical region. They are made up of cases of malaria in children under five (5) years of age, pregnant women and patients aged five and over.

➢ **IT tools used in the study**

Various tools and software were used to conduct, analyse and format the results of this study:
- **R version 4.3.0**

The free version of R Desktop was used for the statistical processing of the study data and the presentation of certain results in graphical form.
- **Microsoft office 2021 package (EXCEL and Word)**

In this study, EXCEL was used to format the climatological and epidemiological data in R format, while Word was used for data entry and text processing.
- **ArcGIS version 10.8**

ArcGIS version 10.8 was used to map the study area.

1.3 Methods

1.3.1 Data processing

Before processing the data, the Jarque-Bera test was performed to check whether the initial data were normally distributed.

Checking the normality of continuous data is a crucial step before carrying out a hypothesis test involving one or more continuous variables. The aim is to ensure that the continuous variables are normally distributed. If this is the case, standard hypothesis tests can be applied. If the normality condition is violated, a so-called 'non-parametric' alternative to the hypothesis test will have to be found.

> **Jarque-Bera test**

The Jarque-Bera normality test is a statistical test that assesses whether a distribution of data is normally distributed on the basis of skewness and kurtosis. It checks whether the data have a skewness and kurtosis similar to a normal distribution (J. B. Cromwell et al., 1994).

> If p-value $>\alpha$: the data follow a normal distribution =H0 ;

> If p-value $< \alpha$: the data do not follow a normal distribution =H1.

The Jarque-Bera test was carried out using R studio software based on meteorological and medical data.

Data processing involves two stages:

1.3.1.1 Descriptive treatment The aim here is to :

- a calculation of monthly averages over eight years for all variables, both meteorological and medical;
- a graphical representation of the average intra-annual variation in the various climatological parameters;
- graphical representations of the average intra-annual and inter-annual variation in malaria, with a view to determining temporal variations and, particularly in cases with large peaks, studying them separately;
- a combined graphical representation of the average intra-annual variation in the number of cases of malaria and that of the various climatological parameters highlighted, with the aim of highlighting the seasons and types of weather favourable to the incidence of malaria.

1.3.1.2 Correlation analysis between each climatological parameter and malaria Simple linear regressions were used to determine which meteorological parameters have a significant influence on malaria. In this case, the relationships were studied individually.

To do this, the correlation coefficient (r) and the P-value were calculated using R Studio software.

➢ Correlation coefficient (r)

The correlation coefficient (Equation 2) measures the strength of the linear relationship between two variables. It is calculated using the following formula:

$$r = \frac{cov(X,Y)}{XY}$$

Equation I-1: correlation coefficient

- Cov (X, Y) denotes the covariance of the variables X and Y ;

- and XY are their respective standard deviations ;

- X represents the explanatory variable (weather parameters) ;

- Y represents the number of cases of malaria.

The correlation coefficient varies between -1 and 1 (Table 3). The closer the value of the correlation between the two variables is to the extremes (-1 or 1), the stronger the correlation (opposite or in phase). Conversely, a correlation of 0 means that the variables studied are linearly independent.

Table 1: Interpretation of correlation coefficients

correlationPositive correlation	Negative
from -0.5 to 0.0 (Low)	from 0.0 to 0.5 (Low)
from -1.0 to -0.5 (Strong)	from 0.5 to 1(Strong)

Source: Sawa, (1978)

➢ P-value

The p-value is used in inferential statistics to conclude on the result of a statistical test. The procedure generally used consists of comparing the p-value with a predefined threshold (usually 5%). If the p-value is below this threshold, the null hypothesis is rejected in favour of the alternative hypothesis, and the test result is declared "statistically significant". Otherwise, if the p-value is above the threshold, the null hypothesis is not rejected, and nothing can be concluded about the hypotheses formulated (Cohen, 1983).

1.3.1.3 Modelling

For this study, multiple linear regression was used to model the seasonal variation in the number of malaria cases.In this case, the statistical relationships between malaria and meteorological parameters are studied in their global interconnections. This approach is the closest to reality, because in nature, the influence of a variable xi is never absolute and individual, but relative and collective (Yaka, 2018) cited by (Passike et al., 2021).Prediction models can be used to estimate the degree of risk expected for a given region at a given time.Modelling involves establishing a cause-and-effect relationship that can be translated into a prediction equation for malaria occurrence as a function of climate factors. The multiple linear regression method was used to generate the model. It is often advisable to use between 60 and 80% of the initial dataset as the training set and the remaining 20 to 40% as the validation set. However, these percentages are not fixed (https://www.aspexit.com).

The data for this study have been divided into two datasets:

- a set of data from January to December of the years 2015 to 2020 called the "training data set" (representing 70% of the initial data) is used to generate the model;

- a dataset called the "validation dataset" running from January to December of the years 2021 and 2022 (representing 30% of the initial data), which was used to check the predictive nature of the generated model, given the short data series.

All the meteorological variables (**Xi**) mentioned above were used as predictors to generate the model.The prediction model obtained is written in the form: **Y=aX1+bX2+...+zXi. Where a, b and z represent coefficients.**

1.3.1.4 Choice of the most relevant variables

The backward elimination method was used to retain the most relevant variables in the model. This method starts with the complete model (all explanatory variables) and at each stage, the variable associated with the highest p-value is eliminated from the model until the remaining variables have p-values smaller than the set threshold (Jolion, 2003). It is important to note that each time a variable is removed, a new multiple linear regression is performed using only the remaining variables.

1.3.1.5 Model validation
Model validation tests

A model can be reliable when :

- the residues are distributed according to the normal distribution ;

- the residuals are linearly independent (no collinearity) ;

- the variance of the residuals is the same for all the values of the explanatory variables (Homoscedasticity);
• no autocorrelation of residuals.

Four tests were used: the Shapiro-Wilk test (W), the Variance Inflation Factor (VIF), the Goldfeld-Quandt test (GQ) and the Durbin Watson test (DW).

➤ **Shapiro-Wilk test**

The Shapiro-Wilk test is used to check whether the residuals in the model follow a normal distribution. It is essential in many areas of statistics because the various tests (VIF, DW, GD) assume that the residuals are normally distributed in order to be valid.
One application of normality tests concerns the residuals of a linear regression model.

- If p-value $< \alpha$: the residuals are not normally distributed;

- If p-value $> \alpha$: the residuals are normally distributed (P. Royston, 1995).

➤ **VIF (Variance Inflation Factor)**

The VIF can be used to check the multicollinearity of the model's dependent variables. Multicollinearity occurs when the model's explanatory variables are highly correlated with each other. The presence of multicollinearity can alter the model's prediction results.
- If VIF<5: No collinearity ;

- If VIF >5, collinearity exists (Akaike, 1974).

➤ **Goldfeld-Quandt :**

The Goldfeld-Quandt test is used to identify the constancy of the error variance. When the variance of the errors is constant, this is referred to as homoscedasticity; when it is not, it is referred to as heteroscedasticity. The homoscedasticity test determines the homogeneity of the variance of the residuals.

- If p-value $< \alpha$: heteroscedasticity of residuals ;

- If p-value $>\alpha$: Homoscedasticity of residuals (Goldfeld and Quandt, 1965).

➢ Durbin-Watson

The Durbin-Watson test is used to detect the first-order autocorrelation of errors in a regression. It assesses whether the residuals of a linear regression model are correlated with each other, which could compromise the validity of the results. This test provides a statistic that takes values between 0 and 4, where a value close to 2 indicates the absence of autocorrelation. If the value is very close to 0 or 4, this indicates strong positive or negative autocorrelation, respectively (Durbin and Watson, 1971).

1.3.1.6 Cross-validation

To ensure that the prediction was accurate, the model to be implemented was validated. The aim was to prove that the model generates good estimates of the values of the variable under study. To do this, it was necessary to work with at least one set of training data and one set of validation data. Quite simply, the training data were used to calibrate the model, while the validation set was used to show that the model is reliable and relevant.

Once the model had been created with the training dataset, it was necessary to calculate objective indicators to assess whether the model generated relevant predictions for the variable under study. The 'true' values of this variable are assumed to be known for all the training and validation datasets. Intuitively, for each sample in the validation dataset, we want to know whether the values predicted by the model are close to the true values in the validation dataset (https://www.aspexit.com).

The indicators used are as follows:

• Coefficient of determination: R2

$$R^2 = 1 - \frac{\sum_{i=1}^{n}(y_i - \hat{y}_i)^2}{\sum_{i=1}^{n}(y_i - \bar{y})^2}$$

(Equation I-2: Formula for R)^2

Where n is the number of measurements, yi is the value of the ith observation in the validation dataset,
$\bar{y}$ is the mean of the values in the validation dataset and $\hat{y}_i$ is the predicted value for the ith observation.
The closer R^2 is to 1, the better the prediction (https://www.aspexit.com).

• The bias

The bias makes it possible to assess whether or not the predictions are accurate and whether the model tends to over- or under-estimate the values of the variable of interest. The bias is calculated as follows:

$$Bias = \frac{\sum_{i=1}^{n}(\hat{y}_i - y_i)}{n}$$

((Equation I-3: Bias formula)

The lower the bias (close to 0), the better the prediction
(https://www.aspexit.com).

• Taylor diagram

The Taylor diagram is a tool for assessing the forecasting model's ability to reproduce actual observations correctly. The closer the points are to the reference line, the more robust and accurate the model is considered to be. This enables more informed forecasting decisions to be made (Taylor, 2001).

CHAPTER II

RESULTS

2.1 Data normality

Table 2: Jarque Bera normality test

Variables	P value	df	X-squared
Umoy	0.51	2	1.3174
Wind	0.60	2	1.0021
Malaria	0.52	2	1.2948
Tn	0.56	2	1.1501
Tx	0.48	2	1.4359
Rain	0.29	2	2.452
INSO	0.29	2	2.4692

Test assumptions :

If p-value >α: the data follow a normal distribution =H0 ;

If p-value < α: the data do not follow a normal distribution =H1.

Given that for all the variables, the P value is greater than the 0.05 threshold, the H0 hypothesis cannot be rejected (Table 2). We can deduce that the data in the study follow a normal distribution.

2.2 Analysis of intra-annual variation in meteorological variables

2.2.1 Analysis of intra-annual variation in precipitation

The rainfall pattern is unimodal, centred on the month of August.

Figure 8 shows the variation in average monthly rainfall between 2015 and 2022. Rainfall is almost zero between November and May, then rises from 0 to 500 mm between May and August and falls from 500 to 0 mm between August and November. Average monthly rainfall is highly variable throughout the year, remaining below 110 mm. Cumulative rainfall during the rainy season (June to October) accounts for more than 98% of the average annual total.

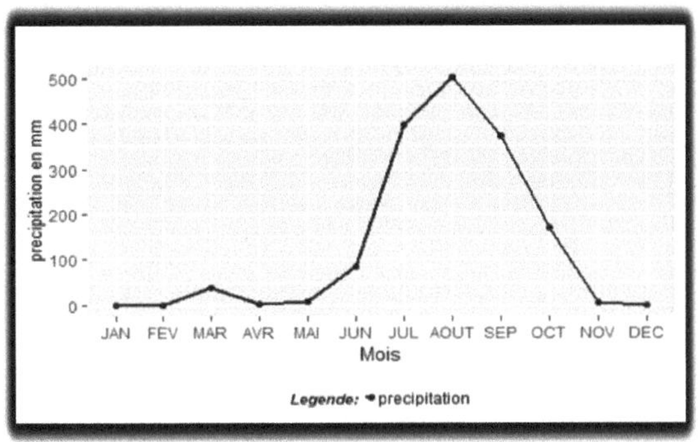

Figure 6: Intra-annual variation in rainfall from 2015 to 2022

2.2.2 Analysis of intra-annual temperature variations (maximum, minimum and average)

There are three periods with contrasting thermal amplitudes: a period with a high thermal amplitude running from January to June, another with a low thermal amplitude running from July to October and a final period with a moderate thermal amplitude running from November to December.

Figure 9 shows the variation in monthly rainfall between 2015 and 2022 in the Ziguinchor region. It shows that :

- maximum temperatures (Tx) fluctuate between 35°C and 41.7°C between January and May, vary between 41 and 32°C between May and August, then remain fairly stable between August and December.

- The average temperature (Tmoy) fluctuates between 24°C and 31°C between January and June, remains fairly stable at 28°C between July and October, then varies from 28°C to 24°C between October and December.

- The minimum temperature (Tn) fluctuates between 14°C and 23°C between January and June, remaining fairly stable at 23°C between June and October, then varying between 23°C and 15°C between October and December.

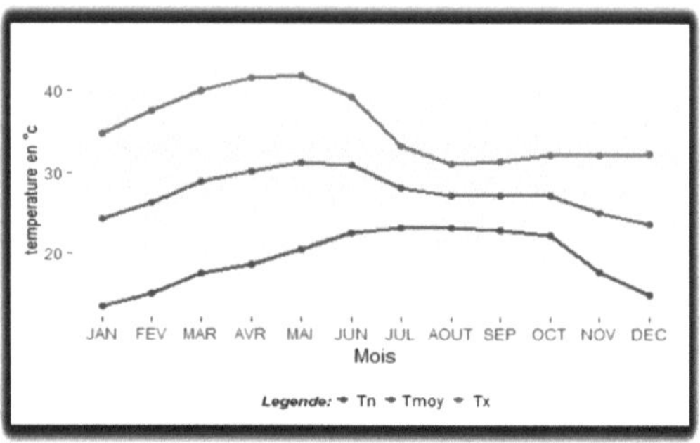

Figure 7: Intra-annual variation in temperatures from 2015 to 2022

2.2.3 Analysis of intra-annual variation in mean relative humidity

Figure 10 shows changes in average monthly relative humidity between 2015 and 2022 in the Ziguinchor region. The analysis shows that average humidity is lowest and stable at 43% between February and March, then varies from 43% to 90% between March and August, then remains stable at 90% between August and September, before falling from 90% to 52% between September and January. Average monthly relative humidity varies considerably over the year, ranging from 43% in February to 90% in August. Humidity therefore appears to be high (>60%) during the period from June to December and low (<60%) during the period from January to May.

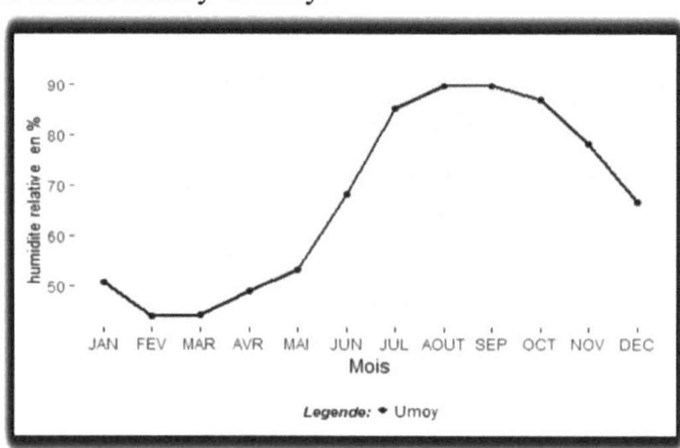

Figure 8: Intra-annual variation in average relative humidity from 2015 to 2022

2.2.4 Analysis of intra-annual wind variation

The average monthly wind speed throughout the year ranges from 0.78 m/s in October to 1.5 m/s in May (Figure 11). Analysis of the intra-annual variation in wind speed between 2025 and 2022 shows that wind speed varies from 0.8 m/s to 1.5 m/s between October and May, then drops from 1.5 m/s to 1.25 m/s between May and July, remains fairly stable at 1.25 m/s between July and August before reaching 0.8 in October. The highest value is observed in May, while the lowest values are observed in October and November. The period of low wind speeds corresponds to the rainy season and the period of high values corresponds to the dry season.

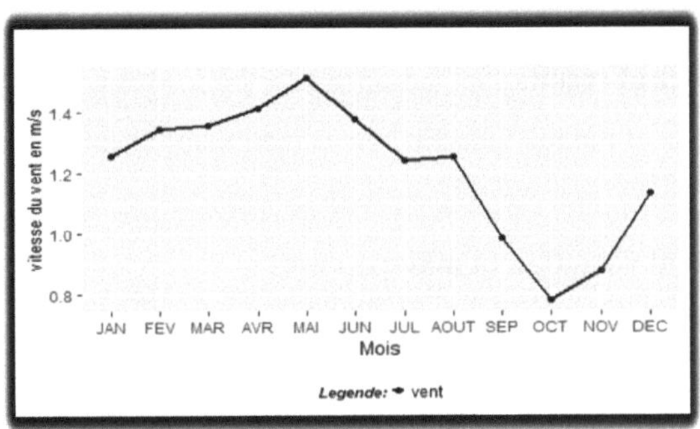

Figure 9: Intra-annual variation in wind speed from 2015 to 2022

• Analysis of intra-annual variation in sunshine duration

We distinguish a period of short duration of insolation due to cloud cover corresponding to the rainy season and a period of long duration of insolation in the dry season. Figure 12 shows the intra-annual variation in sunshine duration between 2015 and 2022 in the Ziguinchor region. Between January and May, the duration of sunshine is more or less constant at 12 hours, then decreases from 12 hours to 8 hours between May and August, then increases from 8 hours to 11 hours between August and November. The average monthly sunshine duration throughout the year is between 8 and 12 hours. The duration of sunshine varies little from November to May and reaches its highest value in May, in contrast to the lowest value observed in August (8 hours).

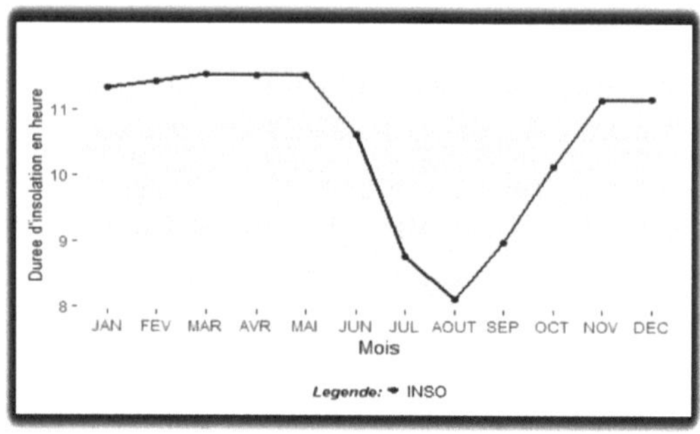

Figure 10: Intra-annual variation in sunshine duration from 2015 to 2022

2.3 Epidemiological results

Graphical analysis of the relationship between climatological parameters and malaria

2.3.1 Inter-annual variation in malaria cases

Figure 13 shows the trend in malaria cases from 2015 to 2022. A decrease is observed, from 8094 to 3319 cases between 2015 and 2018, followed by an increase from 3319 to 7254 cases from 2018 to 2020, and then a significant decrease in 2022. A total of 41,599 cases were recorded over this period, with an annual average of 5,200 cases. The maximum peak is reached in 2015 with 8094 cases, while the minimum is recorded in 2022 with 3145 cases.

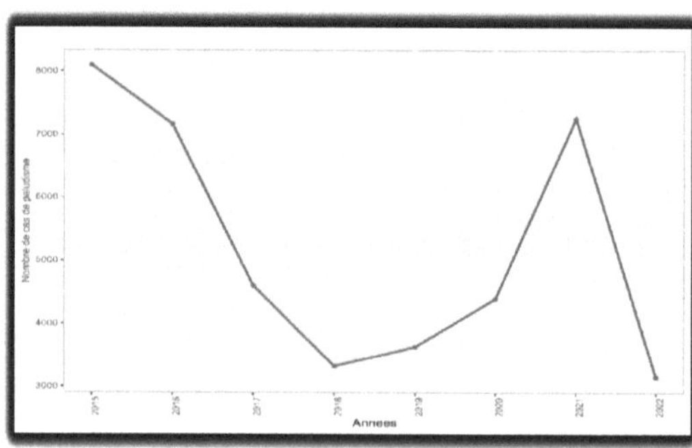

Figure 11: Inter-annual variation in malaria cases from 2015 to 2022

2.3.2 Intra-annual variation in malaria cases

Figure 14 shows the intra-annual variation in the number of malaria cases between 2015 and 2022 in the Ziguinchor region. Between January and February, the number of malaria cases remains constant at 230 cases, then falls from 230 to 120 cases between February and April, then increases from 120 to 300 cases between April and June and remains constant until July before increasing exponentially from 300 to 650 cases between July and October and finally falling from 650 to 350 cases between October and December.

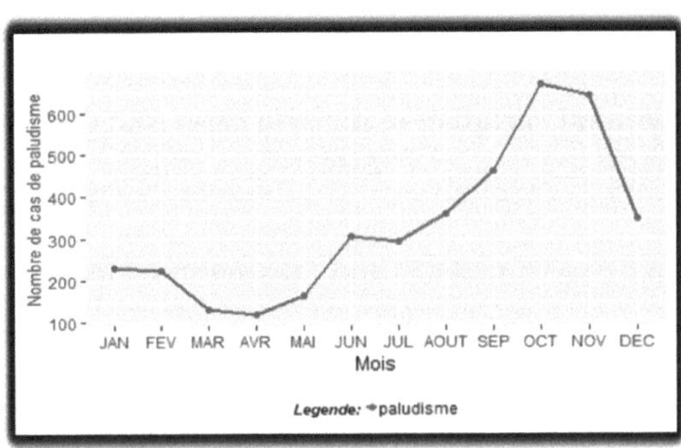

Figure 12: Intra-annual variation in malaria cases from 2015 to 2022

2.3.3 Relationship between malaria and rainfall

Figure 15 illustrates the intra-annual variation in rainfall and the number of cases of malaria between 2015 and 2022. There is a small peak in the number of malaria cases in June which marks the start of the epidemic. This is followed by a rapid increase in malaria cases, which continues until it peaks in October, two months after the peak in rainfall in August. It has to be said that malaria is present throughout the year, except that record numbers of cases are more common in the winter months than in the dry season, although the peak is recorded on average at the end of the winter months.

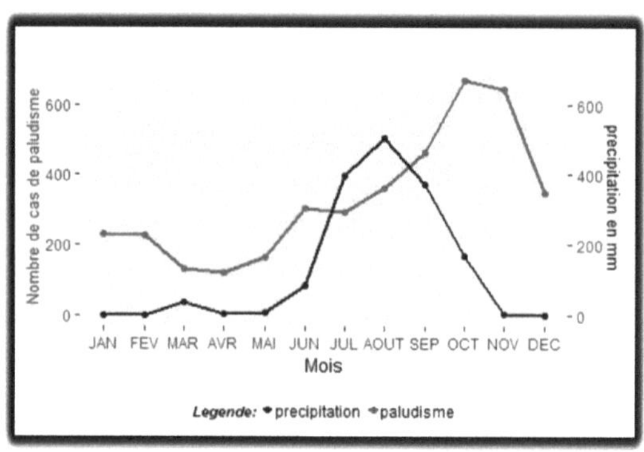

Figure 13: Intra-annual variation in malaria cases and rainfall from 2015 to 2022

2.3.4 Relationship between malaria and maximum temperature (Tx)

The high temperatures are leading to a sharp fall in the number of cases of malaria.Analysis of the occurrence of malaria in relation to maximum temperature shows an inverted trend (Figure 16). During periods (January to June) when the maximum temperature is high, the number of cases appears to be low. In contrast, the number of cases is high during the period of relatively low maximum temperatures (<33.3°C) (July to December).

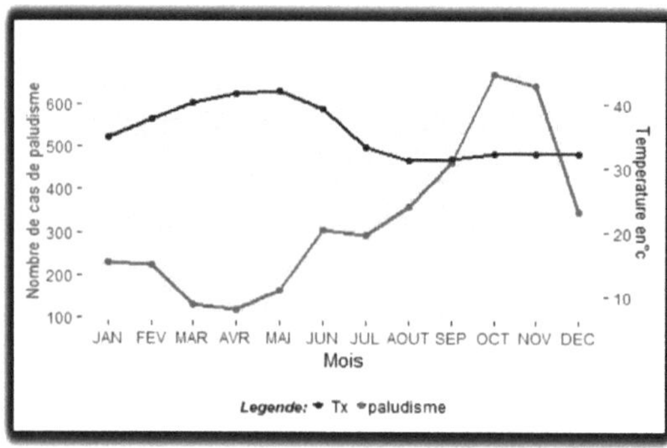

Figure 14: Intra-annual variation in malaria cases and maximum temperature from 2015 to 2022

2.3.5 Relationship between malaria and minimum temperature (Tn)

High minimum temperatures favour an increase in the number of malaria cases. Analysis of the occurrence of malaria in relation to the minimum temperature shows a synchronous trend (Figure 17). Between January and May, when the minimum temperature is low, the number of cases appears to be low. In contrast, the number of cases is high during the period when minimum temperatures are high (>22.5°C) and fairly constant (June to October), then decreases from October onwards, which coincides with the peak in the number of malaria cases, until December with temperatures of 15°C.

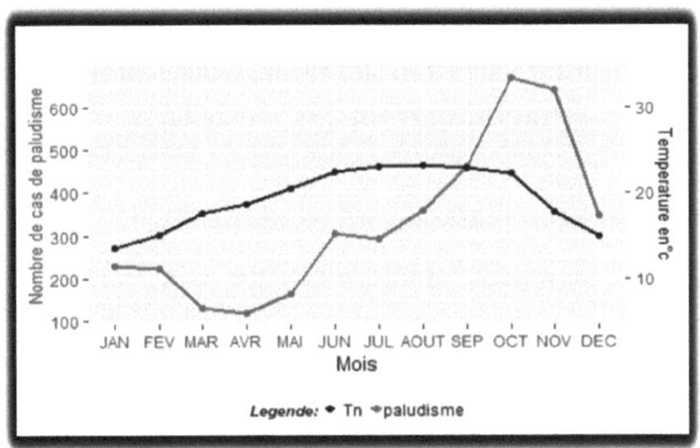

Figure 15: Intra-annual variation in malaria cases and minimum temperature from 2015 to 2022

2.3.6 Relationship between malaria and average relative humidity (Umoy)

High monthly relative humidity is conducive to the occurrence of malaria. Analysis of the occurrence of malaria in relation to relative humidity (Um) reveals that the number of cases of the disease follows the trend of (Umoy) throughout the year. In fact, the increase in the rate of (Umoy) is discreetly accompanied by an increase in the number of cases of malaria from January to June. From July onwards, however, humidity exceeds 85% and varies little at the same time as the number of cases of malaria increases very rapidly, reaching its peak in October. There is a coincidence between the fall in relative humidity and that in the number of cases of malaria from November onwards, when relative humidity falls from 86 to 78.6% (Figure 18).

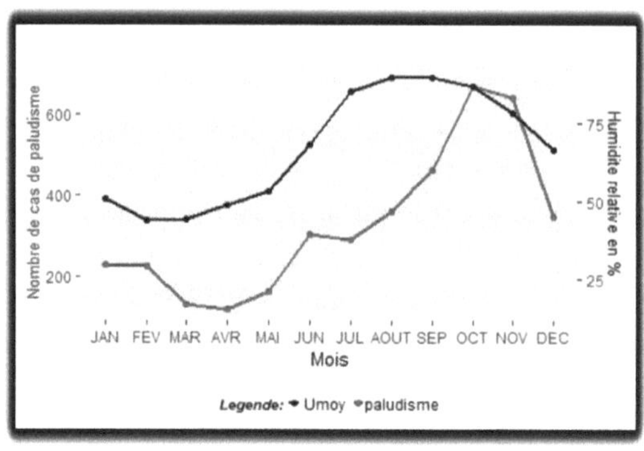

Figure 16: Intra-annual variation in malaria cases and average relative humidity from 2015 to 2022

2.3.7 Relationship between malaria and wind speed

The number of cases increases rapidly when the wind speed drops (<1.3m/s) (Figure 19). It can be seen that when the wind is strong, the number of malaria cases is low, and the opposite is true when the wind becomes weak. The peak in the number of malaria cases coincides with the lowest wind speed (0.78 m/s) of the year in October.

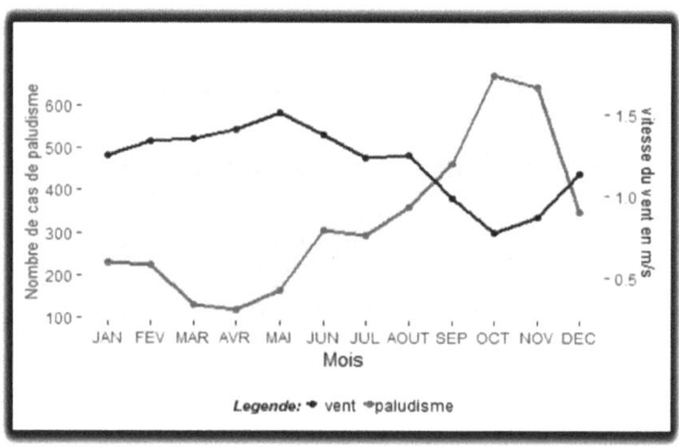

Figure 17: Intra-annual variation in malaria cases and wind speed from 2015 to 2022

2.3.8 Relationship between malaria and sunstroke (INSO)

The length of the day influences the number of monthly malaria cases. Analysis of the incidence of malaria in relation to sunshine duration also reveals a contrasting trend (Figure 20). In fact, the number of cases of malaria has followed the opposite trend to that of sunshine duration. During the periods (January to May and December) when sunshine duration is high and varies very little (11.1 and 11.5 hours), the number of cases appears to be low.

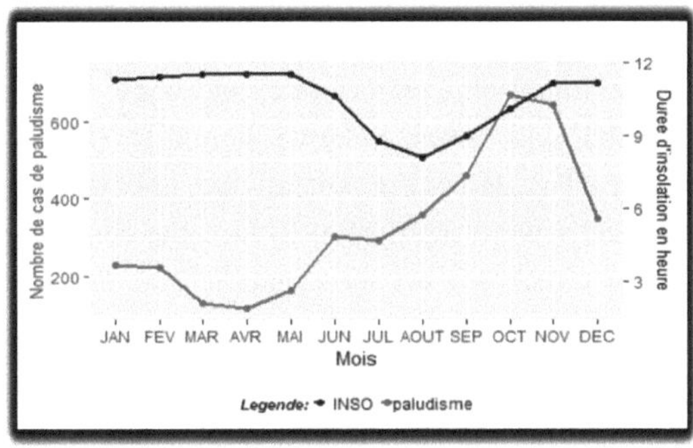

Figure 18: Intra-annual variation in malaria cases and insolation from 2015 to 2022

Generally speaking, the rapid increase in the number of cases of malaria occurs during the period of the year when :
- ➢ the monthly sunshine duration is less than 11 hours;
- ➢ the monthly wind speed is less than 1.3 m/s ;
- ➢ the monthly relative humidity is over 80%;
- ➢ the minimum monthly temperature is above 22.5 °C ;
- ➢ the maximum monthly temperature is below 33.3°C ;
- ➢ monthly rainfall exceeds 175 mm.

These weather conditions occur mainly during July, August, September and October, which suggests that this is the most favourable period for the occurrence of malaria.

2.3.9 Analysis of the correlation between each climatological parameter and malaria.

All the meteorological variables identified appear to be strongly associated with the occurrence of malaria, as the rapid increase in the number of malaria cases coincides with simultaneous variations in all the variables. It is difficult to visually distinguish which variables significantly influence the occurrence of the disease. We therefore applied simple linear regression between each meteorological variable and the number of malaria cases, as shown in the appendix. Taken individually, malaria shows a negative correlation with insolation (INSO), maximum temperature (Tx) and wind. On the other hand, it is positively correlated with mean relative humidity (Umoy), rainfall and minimum temperature (Tn). However, not all the links between malaria and meteorological parameters were significant. In fact, the correlation is strong and significant (at the 5% threshold) between malaria and maximum temperature and mean relative humidity, and highly significant for wind (at the 1% threshold), as shown in Figure 21.

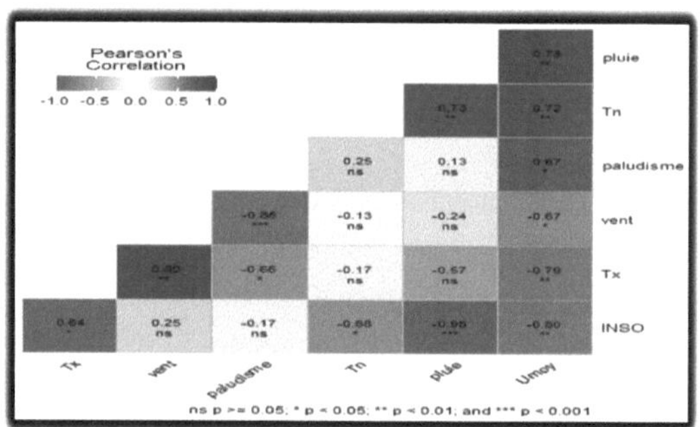

Figure 19: Correlation matrix between meteorological parameters and malaria

2.4 Modelling results

The result of the first multiple linear regression is obtained by taking into account all the available meteorological variables (Table 2). The successive irrelevant meteorological variables are then eliminated using the backward elimination method.

Table 3: Result of the first multiple linear regression

Source	Value	Standard error	t	Pr > |t|	Lower terminal (95%)	Upper terminal (95%)	Meaning codes p-values
Constant	4612,069	3625,976	1,272	0,259	-4708,800	13932,937	°
Tx	-72,820	42,097	-1,730	0,144	-181,033	35,393	°
Tn	92,512	50,194	1,843	0,125	-36,514	221,539	°
Rain	-1,950	1,220	-1,598	0,171	-5,087	1,187	°
Wind	-359,141	352,606	-1,019	0,355	-1265,544	547,262	°
INSO	-168,838	246,475	-0,685	0,524	-802,421	464,745	°
Umoy	-14,927	13,251	-1,126	0,311	-48,991	19,137	°
Significance codes: 0 < *** < 0.001 < ** < 0.01 < * < 0.05 < . < 0.1 < ° < 1							

Number of malaria cases= 4612.06-72.81*Tx+92.51*Tn-1.94*Rain-359.14*Vent- 168.83*INSO-14.92*Umoy (Equation II-1:model 1)

Statistical parameters of model 1 :

To generate the multiple linear regression model, we used all the meteorological variables mentioned above. The interconnection of meteorological variables is strongly correlated with the occurrence of malaria with a highly significant p-value.

Coefficient of determination (R^2): 0.93 R^2 adjusted= 0.86 P. value = 0.008

The coefficients of determination of the model are on the whole greater than 0.7 with a P. Value below the 0.05 threshold.

2.5 Choice of the most relevant variables

We carried out a rigorous selection of variables using the backward elimination method. This approach resulted in the creation of three distinct models, as illustrated in Tables 4, 5 and 6.

Table 4: Results of the second multiple linear regression after elimination of insolation

Source	Value	Standard error	t	Pr > |t|	Lower terminal (95%)	Upper terminal (95%)	Meaning codes p-values
Constant	2287,940	1221,325	1,873	0,110	-700,536	5276,415	°
Tx	-69,596	39,940	-1,743	0,132	-167,325	28,133	°
Tn	75,144	41,358	1,817	0,119	-26,056	176,344	°
Rain	-1,155	0,363	-3,180	0,019	-2,044	-0,266	*
Wind	-155,388	180,792	-0,859	0,423	-597,771	286,994	°
Umoy	-9,153	9,762	-0,938	0,385	-33,039	14,734	°
Significance codes: 0 < *** < 0.001 < ** < 0.01 < * < 0.05 < . < 0.1 < ° < 1							

Number of malaria cases = 2287.93-69.59*Tx+75.14*Tn-1.15*Rain- 155.38*Vent- 9.15*Uaverage (Equation II-2: model 2)

Statistical parameters of model 2 :

In order to develop model 2, all the meteorological variables mentioned above were taken into account, with the exception of insolation. The interconnection of these meteorological variables shows a robust correlation with malaria occurrence, demonstrating a highly significant p-value. However, it should be noted that, with the exception of rainfall, none of the variables showed a significant individual contribution to malaria occurrence (see Table 4).

Coefficient of determination (R^2): 0.93 R^2 adjusted= 0.87P. value = 0.002

The coefficients of determination of the model are on the whole greater than 0.7 with a P. Value below the 0.05 threshold.

Table 5: Results of the third multiple linear regression after eliminating wind

Source	Value	Standard error	t	Pr > |t|	Terminal lower (95%)	Terminal higher (95%)	Codes of significance of p-values
Constant	2635,864	1130,579	2,331	0,053	-37,530	5309,258	.
Tx	-87,367	33,527	-2,606	0,035	-166,647	-8,088	*
Tn	88,975	37,381	2,380	0,049	0,582	177,368	*
Rain	-1,359	0,270	-5,027	0,002	-1,998	-0,720	**
Umoy	-11,317	9,254	-1,223	0,261	-33,199	10,566	°
Significance codes: 0 < *** < 0.001 < ** < 0.01 < * < 0.05 < . < 0.1 < ° < 1							

Number of malaria cases = 2635.86-87.36*Tx+88.97*Tn-1.35*Rainfall- 11.31*AverageU (Equation II-3: model 3)

Statistical parameters of model 3 :

To create model 3, we used the meteorological variables selected to generate model 2, with the exception of wind. The interconnection of these meteorological variables showed a close correlation with the occurrence of malaria, with a highly significant p-value. However, it should be noted that only mean relative humidity did not make a significant individual contribution to the occurrence of malaria (see Table 5).

Coefficient of determination (R^2): 0.92 R^2 adjusted= 0.87 P. value = 0.001

The coefficients of determination of the model are on the whole greater than 0.7 with a P. Value below the 0.05 threshold.

Table 6: Results of the fourth multiple linear regression after elimination of mean relative humidity

Source	Value	Standard error	t	Pr > |t|	Terminal lower (95%)	Terminal higher (95%)	Codes of significance of p-values
Constant	1272,735	194,644	6,539	0,000	823,886	1721,584	***
Tx	-46,914	5,625	-8,340	<0,0001	-59,886	-33,943	***
Tn	44,233	7,897	5,601	0,001	26,022	62,443	***
Rain	-1,108	0,182	-6,101	0,000	-1,527	-0,689	***
Significance codes: 0 < *** < 0.001 < ** < 0.01 < * < 0.05 < . < 0.1 < ° < 1							

Y4=Number of malaria cases = 1272.73-46.91*Tx+44.23*Tn-1.108*Rain
(Equation II-4: model 4)

Statistical parameters of model 4 :

The selected meteorological variables were used, with the exception of mean relative humidity, to develop model 3. The interconnection of these meteorological variables shows a close correlation with the number of malaria cases, with a highly significant p-value. Each of the variables shows a highly significant individual contribution to the occurrence of malaria (see Table 6).

Coefficient of determination (R^2): 0.90 R^2 adjusted= 0.87 P. value = 0.000

The coefficients of determination of the model are on the whole greater than 0.7 with a P. Value below the 0.01 threshold.

Y4 is the final equation of the model after eliminating the least relevant variables.

2.6 Model validation

2.6.1 Validation test results

The Shapiro-Wilk, Durbin-Watson (DW), Goldfeld-Quandt (GD) and VIF tests revealed the following results:

• Shapiro-Wilk test for residuals

The results of the Shapiro-Wilk test (Table 7) confirmed that the residuals of the model follow a normal distribution. This test is essential in many areas of statistics, as the various tests (VIF, DW, GD) assume that the residuals are normally distributed in order to be valid.

Table 7: Test of the normality hypothesis for residuals

W	0,958
p-value (bilateral)	0,755
Alpha	0,05

Test assumptions :
H0: The residuals follow a Normal distribution.
Ha: The residuals do not follow a Normal distribution.
Since the P value is above the 0.05 threshold, the H0 hypothesis cannot be rejected, so the residuals follow the normal distribution.

• VIF (Variance Inflation Factor)

The results of the VIF test verified the multicollinearity of the model's dependent variables. The presence of multicollinearity can alter the model's prediction results.

Table 8: Multicollinearity statistics for variables

	Rain	Tx	Tn
Tolerance	0,261	0,542	0,377
VIF	3,825	1,846	2,651

Test assumptions :
-If VIF<5: No collinearity ;
-If VIF >5, collinearity exists
All the relevant variables have FIVs of less than 5. According to Table 8, there is no multicollinearity between the explanatory variables in the model.

• Goldfeld-Quandt :

The Goldfeld-Quandt test is used to identify the constancy of error variance.

Table 9: Statistical characteristics of the Goldfeld-Quandt test

GQ	1,19
p-value	0,45
Alpha	0,05

Test assumptions :
If p-value >α: Homoscedasticity of residuals =H0; If p-value < α:
Heteroscedasticity of residuals =H1.

Given that the P value is above the 0.05 threshold, the H0 hypothesis cannot be rejected (Table 9). The results show that the variance of the errors is constant - this is known as homoscedasticity.

• Durbin-Watson :

The Durbin-Watson test is used to detect the first-order autocorrelation of errors in a regression. It assesses whether the residuals of a linear regression model are correlated with each other, which could compromise the validity of the results.

Table 10: Statistical characteristics of the Durbin-Watson test

DW	2,16
p-value	0,49
Alpha	0,05

Test assumptions :
If p-value >α: no autocorrelation of residuals =H0; If p-value < α:
autocorrelation of residuals =H1.

Given that the P value is above the 0.05 threshold, the H0 hypothesis cannot be rejected (Table 10). The results of the evaluation of the correlation between the residuals show that there is no autocorrelation between them. The Durbin-Watson (DW) statistic was found to be satisfactory, with a value (DW =2.166) of between 1.5 and 2.5.

2.7 Model robustness

The robustness of a model refers to its ability to perform well when faced with different situations or new data. This often involves rigorous testing on various data sets to assess its stability. The tests are the indicators used to assess the relevance of the predictions of the number of cases of malaria. The indicators calculated from the validation dataset and the model predictions are listed in Table 10.

Table 11: Results of robustness tests on the Y4 model

R^2	0,86
Bias	-60
MAE	129
RMSE	174

Analysis of the Taylor diagram (Figure 22) reveals that the model has poor predictive capacity. It is statistically characterised by a relatively high correlation coefficient (r) of around 91%, an RMSE of 174 malaria cases/month and a standard deviation (σ) of 139 which is very far from that of the observed.

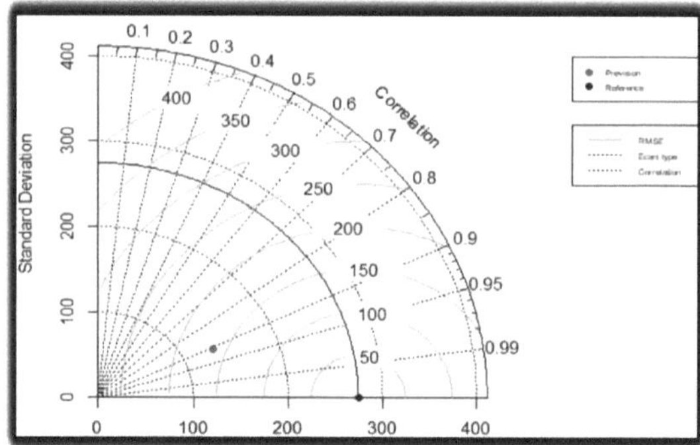

Figure 20: Taylor diagram showing statistical comparisons between predicted and observed values

The model is able to reproduce the monthly trend and seasonality of malaria. Nevertheless, it overestimates (+5%) the number of cases during the dry season and underestimates (-44%) the number of cases during the rainy season (Figure 23). On average, the model underestimates (-18%) the number of annual malaria cases.

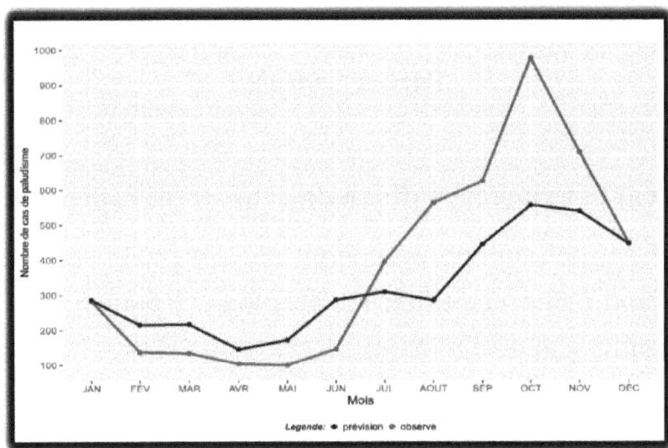

Figure 21: Observed and simulated seasonal representation of malaria occurrence

Given the results of the various statistical tests, we can conclude that the Y4 equation model is valid but not robust.

CHAPTER III
DISCUSSION

3.1 Climatological parameters influencing the occurrence of malaria

The findings of the simple linear regressions clearly establish robust correlations between malaria and some identified meteorological parameters, such as wind, mean relative humidity and maximum temperature. These results reinforce the conclusions o f previous authors, including (Xiao et al., 2020), who argue that epidemiological studies have long suggested that meteorological factors, particularly temperature, humidity and wind, can influence the incidence of infectious diseases. Examination of the various graphs reveals the presence of critical thresholds for relative humidity (85%), maximum temperature (below 33.3°C) and minimum temperature (above 22°C). °C). As soon as these thresholds are reached simultaneously, there is an exponential increase in the number of malaria cases. This suggests that these specific levels of humidity and temperature are optimal for mosquito proliferation, especially as low temperatures induce physiological changes leading to mosquito inactivity, while mild temperatures reactivate their activity (Noubissi, 2021). In addition, temperatures above 40°C are often lethal for mosquitoes (Kirby et al., 2009, cited by Konté, 2021).July, August, September and October seem to offer ideal conditions for mosquito survival, with an optimum combination of humidity and minimum and maximum temperatures. During this period, which is characterised by high rainfall, abundant humidity, short periods of sunshine and low wind speeds, the Ziguinchor region experiences a peak in malaria cases. This increased proliferation of mosquitoes can be explained by the low mortality rate of mosquitoes when both night-time temperature and relative humidity are high (Noubissi, 2021).The length of the aquatic phase varies, from 10 days at a temperature of 30°C (Youmsi et al., 2018) to 15 days at a temperature of 20°C (Coetzee and Fontenille, 2004). This period becomes longer as the temperature decreases and shorter as it increases (Robert, 2001). Thus, during the overwintering period, when minimum temperatures reach their maximum, generations of mosquitoes follow one another at increasingly shorter intervals. This explains the increased proliferation of mosquitoes during the wet season and, potentially, the sharp rise in malaria cases. Although there is a weak correlation between malaria and sunshine, there is also a rapid increase in the number of malaria cases as the duration of sunshine decreases. This trend could be explained by the fact that female Anopheles, responsible for of malaria transmission, prefer to bite mainly at night (dusk to dawn) (Zongo, 2009). As a result, the number of bites increases in proportion to the length of the night, especially as the temperature is generally low during this period.The period of strong increase in the number of malaria cases also corresponds to a period of low wind speeds, as illustrated in Figure 19. This coincidence can be explained by the fact that evaporation rate (mm/day) depends on air temperature, insolation, wind speed and turbulence (Bouzianeb et al., 2015). During this period, puddles experienced reduced evaporation due to moderate air temperature, minimal insolation and low wind speed. The persistence of puddles favours the multiplication of mosquitoes and, consequently, the transmission of malaria. Malaria also persists during the dry season, from December to May, although its incidence is much lower than in the wet season. This reduction can be explained by the fact that, although optimal weather conditions no longer prevail, they still become tolerant to mosquito survival. Maximum temperatures exceed 40°C, minimum temperatures fall below 20°C, and average relative humidity rarely reaches 53%. In these conditions, mosquitoes find it difficult to thrive, resulting in high mosquito mortality. However, they are able to reproduce thanks to the presence of irrigation water used for dry off-season crops such as market gardening and rice growing, as well as the microclimate created by this practice. This explains the weak and non-significant correlation between rainfall and malaria, as shown in Figure 21. In the department of Ziguinchor, market-gardening areas are characterised by non-contiguous housing, with large open spaces in the outlying districts, encouraging interstitial farming around concessions and in empty spaces awaiting construction (S. O. Diedhiou, 2018). Irrigated farming can extend the breeding season, thereby increasing the annual disease transmission period. Irrigation can also increase relative humidity in dry regions, thereby favouring the survival of vectors (Lindsay et al., 2001).

During the dry season, mosquito eggs lie dormant in moist soil to withstand the drought and hatch shortly after becoming sufficiently wet. This explains why Anopheles mosquitoes do not necessarily experience a severe population bottleneck during the dry season, maintaining a large effective population (Minakawa et al., 2005). However, the probability of mosquito survival remains low.In addition, a faulty sewage system and/or the uncontrolled dumping of household wastewater by the local population contribute to the problem. Wastewater and open gutters become breeding sites (breeding grounds) for mosquitoes. The emptied wastewater is often dumped in the surrounding villages (B. Sané, 2017). Surveys conducted by (B. Sané, 2017) reveal that 7.3% of households dump their washing water in the street, 13.0% in the courtyards of houses, 32.7% in vacant lots or unused spaces, and 4.7% in nearby rainwater drainage channels. Permanent breeding sites such as rivers, which contain water all year round, tend to favour malaria transmission throughout the year (Gianotti et al., 2009).During the dry season, wind speed is high, disrupting the stability of mosquitoes in flight and thus having a negative influence on the number of bites and malaria transmission. This explains the highly significant negative correlation between wind and malaria, as illustrated in Figure 20. Previous research suggests that wind can affect the movement of mosquitoes, disrupting breeding sites by drying out aquatic habitats necessary for their life cycle. This disturbance helps to reduce the mosquito population, thereby reducing the risk of malaria transmission.

3.2 Model

The statistical characteristics of the multiple regression of model 1 indicated a strong link between malaria and the interconnection of the meteorological parameters studied. The combined action of the various meteorological parameters has a strong influence on the variation in the number of cases of malaria in the Ziguinchor region. Nevertheless, the search for the most relevant parameters using the top-down elimination method led to the selection of rainfall, maximum temperature (Tx) and minimum temperature (Tn) to generate the prediction model for the number of malaria cases (model 4). At the 5% threshold, the contributions of wind, mean relative humidity and insolation were not significant to the variation in the number of malaria cases. According to the multiple linear regression, the estimated number of malaria cases follows the equation Y4= **1272.73-46.91*Tx+44.23*Tn-1.108*Rain** (Equation III-1: model 4).The model explains a statistically significant and substantial proportion of the variance (R2= 0.90, p < 0.001, adj. R2 = 0.87). The intercept of the model, corresponding to Tx = 0, Tn = 0 and rain = 0, is 1,272.74. In this model :

• The independent variables selected (rain, Tx and Tn) help to explain up to 90% of the variation in the number of cases of malaria in the Ziguinchor region.

• the effect of Tx is statistically significant and negative; when Tx increases by an average of 1 degree, the number of malaria cases falls by an average of 47 units.

• the effect of Tn is statistically significant and positive; when Tn increases by an average of 1°c, the number of malaria cases increases by an average of 44.

• the effect of rainfall is statistically significant and negative; when rainfall increases by an average of 1 mm, the number of malaria cases falls by an average of 1 unit.

From the detailed report of Model 4, it is clear that during the period of November, December, January and February, when the harmattan winds are strong (cool and dry), the minimum temperature (see Figure 24) is the main limiting factor for mosquito development. It has been observed that, in 5 out of 8 years, the peak in the number of malaria cases occurs when the minimum temperature falls from optimal to unfavourable values at the end of the season (Appendix 4). In fact, mosquito mortality increases when night-time temperature and relative humidity are simultaneously low, as indicated by Noubissi (2021). Subsequently, during the period of March, April and May, characterised by strong harmattan winds (hot and very dry), it is the maximum temperature that becomes the limiting factor for mosquito survival (see Figure 24). A temperature rise above 34°C generally has a negative impact on the survival of vectors and parasites, as can be seen from the decrease in malaria cases during the dry season, despite the presence of irrigation water (Rueda, 1990).

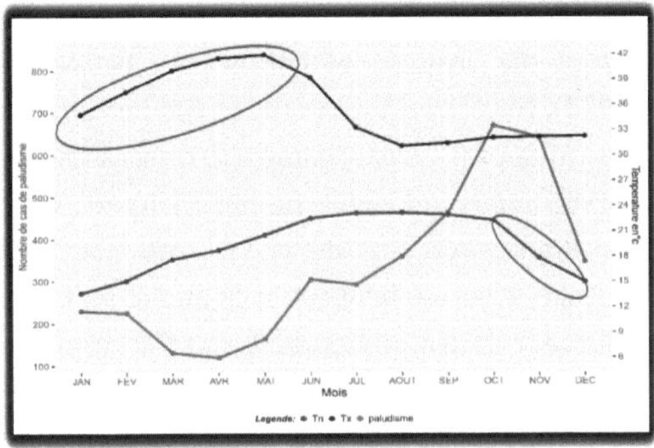

Figure 22: Graphic illustration of temperatures lethal to mosquito survival

The rainfall peak occurs in August, followed two months later by the malaria peak in October. This indicates that the number of malaria cases increases during the period of August, September and October, corresponding to a decrease in rainfall. In fact, low maximum temperatures combined with high minimum temperatures are responsible for the increase in the number of malaria cases (+44 malaria cases/1°C drop in Tx and +47 malaria cases/1°C increase in Tn). The contribution of temperature predominates over that of precipitation due to their high coefficients. This could explain the time lag between the peak in malaria and the peak in rainfall in the Ziguinchor region. It should be noted that during the period from June to August, rain has a reducing influence on the number of cases of malaria (-1 case / increase of 1mm). It is likely that the heavy rainfall in July and August will cause disruption to larval breeding sites through run-off. Excessive rainfall can lead to high water levels, rapid flow velocities and flooding of water reservoirs, which is detrimental to mosquito survival (Martens et al., 1995; Paaijmans et al., 2007), cited by (L. Conté, 2021). Rempel (1953) adds that excessive rainfall can, on the contrary, disturb small breeding sites, destroying the eggs or larvae, as these sites must remain stable from the time the eggs are deposited until the adult emerges. On the other hand, during September and October, rain makes a positive contribution to the number of cases of malaria (+1 case/1mm drop). The breeding sites are less disturbed as the intensity of the rains decreases towards the end of the season. Moderate rainfall is beneficial for immature mosquitoes (water-bound stages) struggling to survive (Martens et al., 1995; Paaijmans et al., 2007), quoted by (L. Conté, 2021). The test designed to evaluate the hypothesis of normality of the residuals indicates that they follow a normal distribution, with a calculated p-value exceeding the alpha significance level of 0.05. Furthermore, the VIF test revealed values of less than 5, suggesting the absence of collinearity between the explanatory variables (Akaike, 1974). Similarly, the results of the homoscedasticity test were satisfactory, with a p-value of 0.13 exceeding the α threshold of 5%. According to Goldef-Quandt (1965), this indicates homogeneity in the variance of the residuals. The Durbin-Watson test also showed the absence of autocorrelation of the residual errors, since a result between 1.5 and 2.5 (Durbin and Watson, 1971) suggests the absence of correlation between the residuals. In the light of these test results, the model generated is considered valid, in accordance with Tomassone et al (1983). Given the large difference between the standard deviations of the forecasts and the observed values, the model is valid but not robust, even though it has a correlation coefficient greater than 0.70 (Figure 22). The model of the dynamics of the number of malaria cases correctly simulates the trend and seasonality of the number of malaria cases over the year (Figure 22). On average, the bias is negative (-18%), indicating a tendency to underestimate the number of annual malaria cases. However, it is important to note that predicted cases are overestimated in the dry season and underestimated in the rainy season. In the dry season, the bias is +5%, while in the wet season it reaches -44%.

3.3 Future malaria trends in Ziguinchor

Given the observation of an upward trend in the minimum temperature (Appendix 2) and a relatively stable one in the maximum temperature (Appendix 1), it is reasonable to expect a significant increase in the number of annual cases of malaria in the future. This could happen if measures are not taken to solve the problems of sanitation and the proximity of dwellings to market gardens and rice farms. In particular, a monthly rise in minimum temperature to 22°C at the end of the wet season, between November and December, would result in thermal conditions favourable to mosquito survival from July to December, thus extending the initial favourable period by two months. An increase in minimum temperatures would have a significant non-linear impact on the extrinsic incubation period (Watts, 1987) and hence on disease transmission.

3.4 Limits of the study

Like most studies of the impact of climate variability on health, this study was limited by the availability and quality of epidemiological data. This meant that we had to conduct the study over a short period of eight years. (08) years, i.e. from 2015 to 2022.

In addition, given that, apart from climatic factors, environmental (physical environment, soil, vegetation, etc.), social, demographic and political factors often play a role in the development of diseases, our study unfortunately did not take these aspects into account. Furthermore, despite the fact that the model reproduced the seasonal evolution of malaria in the Ziguinchor region fairly well, we must bear in mind that the prediction results would include very significant biases.

CONCLUSION AND RECOMMENDATIONS

Malaria, transmitted by the female Anopheles mosquito, has a significant impact on the health of the population, with negative repercussions on the country's economy due to the reduced productivity of a population affected by the disease. The aim of this study is to gain a better understanding of the relationship between climate and malaria, in order to improve the surveillance and management of the disease. It seeks to identify the meteorological characteristics that favour malaria, to assess the statistical links between these parameters and the disease, and to develop a forecasting model based on these relevant meteorological data. The period when malaria is most prevalent is during the wet season, mainly between July and October, when there is an exponential increase in the number of cases. This period is characterised by the onset of the rainy season, an average relative humidity of between 67% and 90%, a minimum temperature above 22.4°C, a maximum temperature below 33.3°C, a wind speed of between 0.78 and 1.5 m/s, and a sunshine duration of between 9.1 and 10.8 hours. Wind and sunshine influence the number of mosquito bites, while temperature, rainfall and relative humidity influence mosquito survival and reproduction. Regression analyses revealed a significant correlation between malaria and maximum temperature, mean relative humidity and wind speed in Ziguinchor. A prediction model based on these variables explains 90% of the variation in the number of cases of malaria, with a high degree of significance for each variable selected. This model is judged to be valid but not robust, offering low-precision predictions with a large error. This study has enriched our understanding of the relationship between malaria and climate. Although wind and relative humidity are strongly correlated with malaria, they are not sufficient to explain the variation in the number of cases in the presence of other variables. The period from July to October remains conducive to malaria because of the favourable conditions for mosquito survival, while the dry season leads to a drop in the number of cases. Minimum and maximum temperatures limit mosquito survival during this period. The model developed in this study could be adapted to other mosquito-borne diseases. However, it does not take into account certain important factors such as irrigation and the promiscuity of neighbourhoods, and underestimates cases during the rainy season and overestimates them during the dry season. To step up the fight against malaria in the Ziguinchor region, we recommend adopting a combined approach, including :

> the preventive use of antimalarial drugs and personal protection measures;

> the introduction of insecticide-treated mosquito nets and the implementation of vector control strategies;

> promoting the draining of marshes or their transformation into flowing water, as well as the elimination of stagnant water points, particularly in the vicinity of dwellings;

> avoid exposure in dark, damp areas during the day in the rainy season;

> avoid building housing in flood-prone or hydro-agricultural areas;

> prevent the uncontrolled dumping of household water (washing powder, dishes) in the streets;

> increase lighting in homes and towns ;

> design homes to allow the sun's rays to reach the rooms ;

> promote air circulation to increase wind speed inside homes.

OUTLOOK

We are well aware that climatic parameters are not the only ones to influence the occurrence of malaria, and it would be wise in the future to integrate social, economic and demographic aspects. Such an approach would be closer to reality, thereby increasing the reliability of the model's forecasts. For the time being, these factors are deliberately left in abeyance, pending further consideration.

BIBLIOGRAPHICAL REFERENCES

Akaïke H., 1974. A new look at the statistical model identification. IEEE Transactions on Automatic Control, 19 p.

A. Kleppe, 2008. Software language engineering: creating domain-specific languages using metamodels. Pearson Education.

A. Pavé, 2012. Modelling living systems From cell to ecosystem. Eco-Energies and Environment.

ANSD, 2019. Situation Économique et Sociale de la Région de Ziguinchor de 2019, P133. **BOUZIANE Brahim, ABID Farid and SAGGAI Sofiane, 2015.** Impacts of water quality on evaporation in an arid environment. P1.

Bouly Sané, 2017. Gestion des eaux usées domestiques et pluviales dans le quartier de Santhiaba-Ouest (commune de Ziguinchor) : Incidences sanitaires et Environnementales, 130P. **Cohen J. and Cohen P., 1983**. Applied Multiple Regression/Correlation Analysis for the Behavioral Sciences. 2nd ed. Lawrence Erlbaum Associates, Inc, New Jersey, 545 p.

D. M. Watts, 1987. Effect of temperature on the vector efficiency of Aedes aegypti for dengue 2 virus. American Journal of Tropical Medicine and Hygiene, 36: p. 143-152.

Durbin J. and Watson G.S., 1971. Testing for serial correlation in least-squares regression, III, Biometrika 58, 1-19.

E. M. Gray and T. J. Bradley, 2005. Physiology of desiccation resistance in anopheles gambiae and anopheles arabiensis. The American journal of tropical medicine and hygiene, vol. 73, no. 3, pp. 553-559.

E. M. Gray, K. A. Rocca, C. Costantini, and N. J. Besansky, 2009. Inversion 2la is associated with enhanced desiccation resistance in anopheles gambiae," Malaria journal, vol. 8, no. 1, pp. 1-12.

E. PASSIKE POKONA1, P. YAKA2, J. A. NDIONE3, 2021. Epidemiological modelling of climate-dependent diseases in northern Togo: the case of foot and mouth disease in the Savanes region Rev. Mar. Sci. Agron. Vét. 9(4) (December 2021) 675-682.

F. Bruneel, 2012. Treatment of severe malaria with intravenous artesunate intravenous artesunate for the treatment of severe malaria, Resuscitation volume 21, pages399-405. **Gbenga J., Abiodun 1 Maharaj R., Witbooi P, Okosun, 2016.** Modelling temperature and rainfall on Anopheles arabiensis population, doi 10.1186/s12936- 0161411-6.

Goldfeld, S.M. and Quandt R.E., 1965. Some tests for homoskedasticity. J.

Am. Stat.Assoc. 60: 539-547.

I. Gaaboub, S. El-Sawaf, and M. El-Latif, 1971. Effect of different relative humidities and temperatures on egg-production and longevity of adults of anopheles (myzomyia) pharoensis theob. Zeitschrift Für Angewandte Entomologie, vol. 67, no. 1-4, pp. 88-94.

Jacques Jouanna, 2020. Climate, environment and health at the dawn of western medicine: Hippocrates [article],sem-link Bulletin de l'Association Guillaume Budé,pp. 28-43.

J. B. Cromwell, W. C. Labys and M. Terraza (1994). Univariate Tests for Time Series Models, Sage, Thousand Oaks, CA, pages 20-22.

Justin-Hervé Noubissi, 2019. Spatio-temporal modelling and simulation of complex dynamic systems with application in epidemiology: the case of malaria. Modelling and simulation. Sorbonne University; Saint Monica University (Buéa, Cameroon), 2019.

Jolion J. M., 2003. Probability and Statistics. Cours de l'INSA. http://rfv.insa-lyon.fr/jolion. Accessed on 23 September 2023.

J. Lourenço and M. Recker, "Natural, persistent oscillations in a spatial multi-strain disease system with application to dengue," PLoS Comput Biol, vol. 9, no. 10, p. e1003308.

J. G. Rempel, 1953. Hatching on moist substrates of the city mosquito (Culicidae). The mosquitoes of Saskatchewan. 44p. 433-509.

J. B. Cromwell, W. C. Labys and M. Terraza (1994). Univariate Tests for Time Series Models, Sage, Thousand Oaks, CA, pages 20-22.

K. Liu, H. Tsujimoto, S.-J. Cha, P. Agre, and J. L. Rasgon, 2011. Aquaporin water channel agaqp1 in the malaria vector mosquito anopheles gambiae during blood feeding and humidity adaptation. Proceedings of the National Academy of Sciences, vol. 108, no. 15, pp. 6062-6066. **Lamine Konté, 2021**. Observation et Modélisation de l'Occurrence du paludisme dans la Région de Ziguinchor (Basse Casamance) avec le Modèle VECTRI, 54P.

L.M. Rueda, 1990. Temperature-dependent development and survival rates of Culex quinquefasciatus and Aedes aegypti (Diptera: Culicidae). Journal of Medical Entomology,. 27:
p. 892-898.

L.M.Rueda, 1990. Temperature-dependent development and survival rates of Culex quinquefasciatus and Aedes aegypti (Diptera: Culicidae). Journal of Medical Entomology,. 27:
p. 892-898.

L. Martin, 2020. Malaria in the Niakhar area of Senegal: mortality and risk factors. Study of malaria mortality in children under 15 and its determinants in rural areas.

Lindsay S.W., Birley, 1996. Climate change and malaria transmission. Ann. Trop. Med. Parasitol, 6,573-588.

M. N. Bayoh, 2001. Studies on the development and survival of Anopheles gambiae sensu stricto at various temperatures and relative humidities. PhD thesis, Durham University.

M.-H. Wang, O. Marinotti, A. Vardo-Zalik, R. Boparai, and G. Yan, 2011. Genome-wide transcriptional analysis of genes associated with acute desiccation stress in anopheles gambiae. PloS one, vol. 6, no. 10, p. e26011.

M.Gillies, 1953. The duration of the gonotrophic cycle in Anopheles gambiae and An. funestus with a note on the efficiency of hand catching. East African Medical Journal,. 30: p. 129-135. **Minakawa N, Munga S, Atieli F, Mushinzimana E, Zhou GF, Githeko AK, 2005**. Spatial distribution of anopheline larval habitats in Western Kenyan highlands: Effects of land cover types and topography. Am J Trop Med Hyg. 2005;73:157-65.

O. Faye', D. Fontenille2, J.P. herve3, P.A. Diack4, S. Diall05 & J. Mouchet, 1933. le paludisme en zone sahelienne du senegal. Ann. Soc..be/ge Méd. tro ', 73, 21-30.

WHO, 2019. "The World Malaria Report 2019," WHO.

Oscar Assoumou Menye, Fabrice Arnaud Guetsop Sateu, 2017. L'entrepreneuriat féminin au Cameroun : enjeux et perspectives Revue Congolaise de Gestion, vol.2, pp 11 à 42.

P. Bourée, 2006. Revue Francophone des Laboratoires Volume 2006, Issue 385, Pages 25-38.

PNLP, 2019. Malaria statistics_ National Malaria Control Programme.

P. Cailly, 2011. Modelling the spatio-temporal dynamics of a population o f mosquitoes, sources of nuisance and vectors of pathogens.

Pascal Zongo, 2009. Modélisation mathématique de la dynamique de transmission du paludisme, PhD thesis, University of Ouagadougou,144P.

Lindsay, S.W. and M.H. Birley, 1996. Climate change and malaria transmission. Annals of Tropical Medicine and Parasitology,1996. 90: p. 573-588.

Patrick Royston, 1995. Remark AS R94: A remark on Algorithm AS 181: The W test for normality. Applied Statistics, 44, 547-551. doi:10.2307/2986146.

R. Kühne, 1975. Fixpunkte positiver operatoren in kb-räumen. Mathematische Nachrichten, vol. 65, no. 1, pp. 259-280.

René Joly Assako Assako, Daniel Bley, Frédéric Simard, 2005. Apports des sciences sociales et de l'entomologie dans l'analyse de l'endémicité du paludisme à HEVECAM, une agro-industrie du Sud-Cameroun Geo-Eco-Trop, 101-114.

S. DOUMBIA, 2010. Impact of climate change on the incidence of malaria in Mali. Thesis of the Faculty of Medicine, Pharmacy and Odonto-stomatology.

S.-L. Claudine and P. Alain, 2002. Environnement : modélisation et modèles pour comprendre, agir ou décider dans un contexte interdisciplinaire. Natures Sciences Sociétés, vol. 10, pp. 525.

S. Diop a, M. Ndiaye a, M. Seck a, B. Chevalier b, R. Jambou c, A. Sarr d, T.N. Dièye a,

A.O. Touré a, D. Thiam a, L. Diakhaté, 2009. Prevention of post-transfusion malaria in endemic areas. Prevention of transfusion transmitted malaria in endemic area Author links open overlay panel a Transfusion Clinique et Biologique Volume 16, Issues 5-6, Pages 454- 459.

SC. F. Darriet, R. N'guessan, A. A. Koffi, L. Konan, J.M.C. Doannio, F. Chandre & P. Carnevale, 1999. Impact de la résistance aux pyréthrinoïdes sur l'efficacité des moustiquaires imprégnées dans la prévention du paludisme : résultats des essais en cases expérimentales avec la deltaméthrine Institut Pierre Richet, OCCGE, Manuscrit n°2133.

Sécou Omar Diedhiou, Oumar Sy and Christine Margetic, 2018. Agriculture urbaine à Ziguinchor (Sénégal) : des pratiques d'autoconsommation favorables à l'essor de filières d'approvisionnement urbaines durables, 40. https://doi.org/10.4000/eps.8250.

T. K. Yamana and E. A. Eltahir, 2013. Incorporating the effects of humidity in a mechanistic model of anopheles gambiae mosquito population dynamics in the Sahel region of Africa. Parasit Vectors, vol. 6, p. 235, 2013.

V. Robert, H. Dieng, L. Lochouarn, S. F. Traoré, J.-F. Trape, F. Simondon, D. Fontenille, 1998. La transmission du paludisme dans la zone de Niakhar, Sénégal Volume3, Issue8, Pages 667-677.

Xiao Y. et al., 2018. The influence of meteorological factors on the incidence of tuberculosis in southwest China from 2006 to 2015. P28.

WEBOGRAPHY

https://www.aspexit.com/comment-valider-un-modele-de- prediction/#Cross-validation_proceduresRetrieved from the 23 September 2023

https://www.who.int/fr/news-room/fact-sheets/detail/malaria Accessed on 23 September 2023